Animals Love Just Like Us

by Scott Marshall
Author of *Love, Explained*

For Penelope

Copyright © 2019 by Scott Marshall

All rights reserved. This book or any portion thereof may not be reproduced or used in any manner whatsoever without the express written permission of the author except for the use of brief quotations in a book review.

First U.S. Edition 2019

V 1.2   03-21-2019

ISBN: 978-0-9994506-3-5 (hardcover)

Library of Congress Control Number: 1-7312122801

With thanks to Jon Goldstein, the first to hear the idea for this book, and Kip Rosser, the first to see a sample mockup, for their insightful suggestions and encouragement, and faithful supporters Stephen, Bertrand, and Valerie.

Contents

The World of Animal Love   2

Lessons of Love From Science   28

Nature's Circle of Love   29

Love is what nature gave us so we will like each other, work together, and not fight or be afraid of each other. It can be everyday love like "let's play together." It can be the very special love between you and your mom or dad, or between girls and boys who want to be very, very close. Love feels good because nature is telling us it's better to be friends than to be enemies.

Feelings are nature's way to tell us what's good for us and what's not good for us. It feels good to have friends and family we like and trust. It feels bad to be alone and think no one is on our side. One of the best feelings is to be part of a group where we all like each other.

You will love someone who makes you feel good, like hugging, caressing, grooming, or saying nice things.

Some animals, like monkeys, like to groom each other to show they are friends or want to be friends. Grooming is taking care of someone's hair or body. It feels very good to be touched by someone you trust, like your mother or best friend, especially if they are taking care of you, like cleaning, fixing your hair, or putting a bandage on a cut or scrape. All these are acts of love.

Family smells good to us, and we smell good to family and friends when we are clean and healthy. The nose knows how important it is that we like the smell of who we love.

Comfortable, caring hugs and caresses make love happen. It's a good idea to see that the one you love likes the way you touch them. You can learn the best way to touch the ones you love by noticing when it makes them feel good.

You know how we show those we love that we trust them? By showing them we know they will not hurt us. Cats like to show this by closing their eyes or lying on their backs. It means "I trust you," which means almost the same as "I love you."

What's a dinosaur doing in this book? Babies have to trust their mother after they hatch. Here is a nest of real dinosaur eggs many millions of years old! See how carefully the mother placed them in her nest? We think she protected them from other animals who might have wanted to eat them. The painting on the other page is what a dinosaur mother guarding her eggs may have looked like. Did mother and baby dinosaurs love each other? We don't know if they could feel love, but they probably acted like dinosaur mothers and babies loved each other very much.

Nature made us to be together with friends and family. We worry when we are alone. It is harder for us to stay healthy when we worry. When we are sick, our bodies need to fight to make us well. If we are alone and away from family, our bodies are busy fighting to not be alone when they need that energy to fight to get well.

Hospitals sometimes help sick children to become healthy by giving them dogs they can cuddle so they will not feel so alone.

If you get sick, what kind of animal would you like to cuddle to help you get better?

Tickling is a way to tell someone you like them and want them to be happy. It's great to be tickled by someone you like. Tickling is a kind of safe touching that the body needs. Did you know that many animals like to be tickled? Even little animals like mice love to be tickled, but their laughing is so high and soft that we need special microphones to hear it.

Hugging can be like *Goldilocks and the Three Bears*, the story where the girl tried sleeping in a bed that was too hard, one that was too soft, and another that was just right. If you love someone and want to hug them, you have to hug them just right—not too hard, and not too soft.

How do you know if you are hugging just right? Pretend you are the person you are hugging and decide if you would like how you are being hugged. Watch how they like your hug, and think about how they feel when you hug them.

Some animals like to be alone most of their lives. The opossum is one. Nature made it so it doesn't get lonely unless it is so young it needs its mother, or it has grown up and needs to find a mate to make babies with, or becomes a mother and needs to take care of her babies. Those are the times when an opossum does not want to be alone.

Some people also don't mind being alone, and that's fine if that's what they really want. Nobody should force you if you really want to be by yourself, but it's real nice if there's someone to be with when you are all done being alone. Everybody needs time alone sometimes, and friends should respect it. Being alone when you want is OK.

Storybook King Arthur was taught life lessons by Merlin the Magician. When Arthur was a boy, Merlin turned him into many different animals, one a day, so he would learn how wise the animals were. His knights wore armor, like here, because they were afraid to be hurt by each other.

When Arthur grew up and became King, he wanted to make everyone stop fighting and be together as friends.

What animal would you like to be for a day?

We feel good when we do something nice for someone. If the person we are nice to is a nice person, then they will do something nice back to us when they can. It's good for you to be nice to those you love when they love you back.

Maybe you can show someone your love right now!

Lessons of Love from Science

What is love?
Love is what nature gave us to help decide who's our friend, who we want to be friends with, and who we can trust to be good for us. When we feel good about someone, our bodies make a certain chemical inside called oxytocin (ock-see-TOE-sin). It changes how we act toward people we feel are good for us. Some scientists call it the love hormone*.

Do animals love?
All animals need to decide who is a friend and who is not a friend. Some animals like to be alone, except when they are briefly in a couple to make babies, or when taking care of babies until they are grown up enough to be on their own. Other animals, like us, are born to be in groups all their lives and are almost never happy alone.

What can we do if we want someone to love us?
We have to let them know that we are on their side, that we are good for them and not bad for them, by what we say to them and what we do to them. When we make them feel good and not feel bad we let them know we are their friend.

*A hormone is a chemical that sends a message from one part of your body to another.

Nature's Circle of Love

When a mother pushes her baby out, if everything follows nature's plan, her body makes a love hormone* so she will feel good about her new baby and want to keep the baby close. It makes her want to cuddle and hug her baby. The baby wants to be with someone caring, which may be the baby's mother, father, sister or brother, a grandparent, or other family or friends. The mother also wants to feel the baby drinking milk from her. That makes more love hormone in the mother's body, and that makes her love her baby even more. The warm, sweet milk the baby tastes helps the baby feel good about the mother, and they become closer and closer and love each other more and more.

The baby grows up and, after being raised with family and friends full of love and cuddling, some day finds someone for romance. Then, they become close and trusting enough together to have their own baby.

This is life's circle of love for us and much of the animal world.

*The oxytocin chemical is the body's love hormone.

Photo Credits

Cover: Pandas, zeleno (iStock)
Title Page: Swans, Freeprod (Onepixel)
Copyright Page: Dachshund, Anthony Maina (Unsplash)
Contents Page: Horses, delfi de la Rua (Unsplash)
2: Lion and Lamb (Getty Images)
4: Lionesses, Joel Herzog (Unsplash)
6: Monkeys, Andrea Izzotti (Onepixel)
8: Seals, boryanam (Adobe Stock)
10: Cats, idmanjoe (Adobe Stock)
12: Kitten in Lap, serezniy (123RF)
14: T-Rex, Elenarts (Adobe Stock)
15: Dinosaur eggs, Maksim Shchur (Adobe Stock)
16: Therapy Dog, Monkeybusinessimages (Onepixel)
18: Mouse: asife (Adobe Stock)
20: Lion: kitkorzun (Adobe Stock)
22: Opossum, Hkuchera (Adobe Stock)
24: Dog and Cat, Krista Mangulsone (Unsplash)
25: Knight, Andrii Podilnyk (Unsplash)
26: Giraffes, John Michael Vosloo (Shutterstock)
27: Hug, Xavier Mouton Photographie (Unsplash)
28: Birds, Roi Dimor (Unsplash
29: Mother and Baby, Anneka (Shutterstock)
30: Family, Laercio Cavalcanti (Unsplash)
Photo Credits Page: Friends, Duy Pham (Unsplash)

www.ingramcontent.com/pod-product-compliance
Lightning Source LLC
Chambersburg PA
CBHW041420160426

42811CB00104B/1836